THE PIT CHILDREN

by

ERIC FORSTER

Northern History Booklet No. 83

I.S.B.N. 0 85983 104 3

To Elaine, Alastair and Andy

Published by Frank Graham, 6 Queens Terrace, Newcastle upon Tyne NE2 2PL
Printed by J. & P. Bealls Ltd., Gallowgate, Newcastle upon Tyne NE1 4SA

FOREWORD

This book has its birth in a series of articles written for the Northern newspaper, the *Sunday Sun*. The author was subsequently approached by several interested readers who were kind enough to suggest that the articles were worthy of expansion and presentation in book form. I was coincidently approached by publisher Mr. Frank Graham with a similar suggestion. The result – much delayed – is herewith presented.

It remains to express sincere thanks to Mr. Charles Parish, librarian of the Literary and Philosophical Society, of Newcastle, and his staff, for their unfailing help and generosity in allowing this member to remove and retain beyond the allotted period those books essential to research.

Thanks, too, to the staff of the Newcastle Central City Library not only for searching out some of the more obscure titles in their collection, but for allowing me research facilities on the premises.

Errors, where they exist, are solely mine. The hope is that the book will make some useful, if small, contribution to the region's ever-growing library of local history. The story told is part of us.

Mayo
Hookergate
County Durham
1978.

CHAPTER ONE

"This nation is in duty bound,
To prize those who work underground,
For 'tis well-known this country round,
Is kept up by the coal trade. . ."

— The Coal Trade.

It was an age which produced before the eyes of a genteel, Victorian middle-class such evidence as this, from a Nottingham master chimney-sweep: "The children suffer terribly and the cruelty they have to endure is quite shocking. They have first to have their elbows and knees rubbed in strong brine, so as to harden the flesh. They are then put to their work and I have myself seen them return with the skin hanging in flaps from their knees. The wounded flesh must again be well rubbed in brine. The little ones do scream so, but it is the only way to harden them to their work and they get used to it. . . ."

He was referring, of course, to the scandal of Britain's boy chimney sweeps, those infants of six years of age and upwards, whose frailty and diminished stature – the physique of undernourishment – made them suitable fodder to be thrust into the home chimneys of the rich of the age.

There was a distinct malevolence about attitudes towards child labour in those early years of Queen Victoria's reign, when the thump and the thunder of the steam age was being heard throughout the land, and the national pursuit of Empire and conquest was about to reach its zenith.

The young Queen on the throne had yet to achieve the deified status that came to her after she donned widow's weeds, and made her in later years the "Mother of the World," yet the portents were there.

True, there were difficulties to be overcome, but neither Rome nor the British Empire was built in a day and who could hold it otherwise than that Britain, home of democratic and individual freedom, would produce a type of world rule which would prove infinitely more durable than that of Rome's?

Laisser-faire economics held full sway; the creed that an employer, be he Lancashire millowner or Durham coalowner, was responsible first to the welfare of his business and then to the wellbeing of his employees, was the hinge upon which swung the potential might of the nation. The British ruling class was always just; the British ruling class would see justice done to Britons. . . .

The accounts books of merchants and factory owners, shippers, ironmasters and coalowners began to be imbued with almost mythological qualities and there were indeed those employers who, as a result, undertook wider social responsibilities towards the men they employed.

But as the blunt wedge of industry was thrust deeper yet into what, in the early 1800's, was still principally an agricultural economy, so there came an abuse on a scale which Englishmen of previous ages would have fought to condemn. The old order changed, and it changed so rapidly that apple-cheeked, cider-fed yokels were transformed into wan-complexioned, soup-begging operatives within the twinkling of an industrial entrepreneur's eye.

New philosophies were born to suit the mood of the day. "Where there's muck there's money," or "look after the pence and the pounds will look after themselves," were two of the axioms parroted by the North's new, pot-bellied rich 'ere they swung gold watches into weskit pockets and ordered buzzers blown to recall their men to twelve hours labour (or even sixteen or eighteen hours).

A new breed of Englishman emerged from the womb of the industrial revolution; one who fitted neatly into the lower levels in the stratification of society that was taking place. The farming hind became the cotton mill operative, the shepherd came down from the hills to become the coal miner, the charcoal burner left the forest glades and emerged out of chrysalis to become a fully-fledged navvy. The new breed was called the "workingman" and they were quickly made aware of their low status in the order of things Victorian.

In no part of the nation was the new order made more manifest than in the North. Here was the spawning ground of industrial inventiveness; here was where men performed the new economic miracle; this was where the coal was mined to fire the boilers which raised the steam to drive the engines which produced the machines and the goods which, in turn, brought home the riches.

And the riches, in turn, worked yet new wonders. Industrial towns fattened on the bones of what once had been rural hamlets. New towns – ugly grid-iron, red-brick ant heaps – sprouted out of the countryside, rooting out not only the few sparse, native inhabitants, but also an entire way of English life.

Ugliness battened on beauty; there was a distortion of things spiritual, temporal and humanitarian ("the economic health of the nation depends on child employment in the mines," said Lord Londonderry).

And in that last phrase was capsuled all that the reformers of the early and mid-Victorian period saw was wrong with the policy of laisser-faire economics. They were few, but they were also influential and they became increasingly vociferous as the century progressed.

The reforming zeal of the few was not, however, worked within a vacuum, for there were other forces at work. They swam beneath the surface of life in the early years of the century and surfaced with such as the reform movement and, later, with "physical force" Chartism. Revolution was never far from the minds of rulers who had the lessons of the French upheaval from which to draw conclusions.

The abuse of child labour was not, however, one of the major motivational forces of the revolutionary reformers, but it was certainly a consideration, as revealed by the "fighting charter" adopted by Northumberland and Durham miners when they flouted authority by forming a union. This was the "Pitmen's Union of Tyne and Wear," which was hardly out of gestation before a strike took place.

One of the results was agreement by mineowners for a twelve hour working day for boys, instead of one without limitation. The year was 1831, and the concession was deemed a great victory but, with the union "broken" by a second strike the following year, the shifts worked underground by children aged six (and on occasion even younger) once again became stretched to the more usual sixteen hours.

More than a decade passed before the plight of the pit children was impartially examined. And then there came a public upheaval of such virulence that it left even the hardened mineowners bruised by reaction. But not for long. . . .

The children of the cotton mills of Lancashire had previously achieved emancipation, of a sort, through the efforts of parliamentary reformers. The child chimney sweeps found public sympathy when their plight was made known. But the pit children worked on unknown; performing work which left them physically stunted for life, in conditions later described as "not better than sewers," and for pittance wages of ten pence per day.

They worked six days a week without seeing daylight. They were carried to work still asleep on the shoulders of fathers or brothers. Below ground, they sat in darkness by their "trapdoors"* through the loneliness of a sixteen hour and, on occasion, eighteen hour shift, awaiting the rumble of coal tubs which was their signal to open the door.

They had no education beyond that afforded by the village Methodist Sunday School, which even if they attended was at best sketchy.

They were indeed children of the pit, bound to it for the remainder of their lives or until creaking age intervened and the poorhouse beckoned.

The South knew nothing of them; those who directed the burgeoning economic fortunes of the North cared nothing.

The revelations of abuse, when finally made nationally known in 1842, revealed another England to those outside the coalfields, an England of savagery and brutality and exploitation which recoiled upon those of genteel instinct and reforming zeal. It also came at an inopportune time for those who at that time were sensing the undercurrents of revolt threading through society. It was a real enough threat. Just four years previously, in May, 1838, Chartism gave birth to the first, specific, working-class movement in Britain. It was then that the *People's Charter* was published. Chartist leaders, declaring that the Reform Act of 1832 fell far short of ideals, sounded a more emphatic rallying cry. For the nation's rulers, the six points of the new charter read more like the trumpetings of revolution, with its demands

*Trapdoors of wooden construction were sited at strategic points throughout the mine galleries in order to direct or "course" air into the workplaces. The doors were opened only to allow the passage of coal tubs pushed by "putters" (usually older boys) or pulled by horses directed by a "driver" (a position of promotion for the trapper boys).

for universal suffrage, vote by ballot, payments of Members of Parliament, no property qualifications, and annually elected Parliaments.

It was not as though the Charter was the work of isolated idealists. It held rich appeal to the mass of the population, the unlettered and unwashed, who felt oppressed by the Corn Laws and frustrated by the lack of involvement in the ordering of their own affairs. It was perhaps the darkest period of English social history – "a knife and fork question" said Carlyle. Hunger was rife, unemployment officially decreed as "disastrous" and existing laws brought not relief to the situation, but an exacerbation of existing conditions. It was the age of the Tolpuddle Martyrs, "peelers" rick-burning, of "dark, satanic mills," of harshening poor laws, of Old Sarum politics, Corn Laws and starvation.

Concessions were made and were aimed at allaying the discontent in the new industrial towns and, thus, legislation came to alleviate the plight of those embroiled in the factory system, and – specifically – to aid those employees of tender years or female sex.

In 1802 there had been passed the "Health and Morals of Apprentices Act," which limited hours of work to twelve per day, curtailed night working, and required masters to allow their apprentices schooling facilities. By 1833 there was need of the "Factories Act" which further limited the hours of work for children to nine hours a day, extended the regulations to all the textile industries and – importantly – provided a scheme of national inspection. Active in the promotion of this act was one whose name was to re-occur in a similar connection a decade later – Lord Shaftesbury.

In the context of its time the Act seemed to be enlightened social legislation and those responsible could sit back and reflect on their efforts with a sense of satisfaction.

But what of the coal mines? What had happened to the infants and girl children employed in the collieries of the North, Wales and Scotland? Was their lot such that reformists could ignore them and turn elsewhere to apply their zeal for social betterment? Why was it deemed unnecessary to extend the Factories Act to include the children of the pits?

The answer, simply, was that no-one outside the industry knew what was happening down there beneath the fifty fathom line. The coalfields were geologically estranged from the rest of Britain. They were peopled by those who spoke a different tongue and whose way of life seemed peculiar and alien to the "outside" world. Theirs was a remote world of clanking steam engines, smoke, rattling wagons, belching fires, noise and thundering machines. It was a world understood only by those who were part of it.

There was recognition, of course, of the worth of coal to the national economy; indeed it was the fulcrum of Britain's prosperity. Coal not only powered the machines of the factories from whence emerged the stuff of Empire, it also – according to one Victorian writer – helped Britain to "supply the very necessities of mankind." Britain, it was asserted, owed its world supremacy to coal. "Let civilised nations be deprived of their coal, and it is difficult to see how they could hold together," he said.

The coal mines of Wales, and Scotland, but above all those of the North, supplied the basic commodity of world wealth. Britain, in the 1840's, was giving the world 31,500,000 tons of coal annually. Belgium, its nearest competitor, could produce only five million tons. The vast coalfields of America were only just being opened up and could produce only four and a half million tons.

Thus, Britain powered the age of industry, and great was the indignation when it came to light that it was being achieved at terrifying cost to generations of its children.

Great, also, was the power of reaction. . . .

CHAPTER TWO

"There is me horse, an' there is me tram;
'Twe horns full of greese will mek her to gan;
There is me hoggers likewise me half shoon,
An' smash me heart marra, maw puttin's a' done!"

— The Collier's Rant.

On 18th June, 1839, the St. Hilda Colliery, at South Shields, blew up. The explosion claimed fifty-two lives which, by the standards of the times, was not unduly remarkable, but there were other factors present which were to combine to make the event notable well beyond the Northern coalfield. Among those who immediately went underground to assist with rescue work was an indefatigable townsman named James Mather, and, following his exertions, there was formed the "South Shields Committee to Investigate the Cause of Accidents in Coal Mines." The committee attracted the attention of some of the mining industry's leading figures and their report, the culmination of three years work, proved the most thorough and scientific treatment of the subject ever undertaken. They examined ventilation of mines, the efficiency of safety lamps, the need for Government inspection of mines and the provision of medical treatment after explosions. They also investigated – and subsequently deprecated – the use of infant labour in the mines.

It was noted that of the fifty-two victims of the St. Hilda explosion, there were six boys under twelve years of age. They reported: "The Committee have assigned reasons for the opinion that admitting children . . . at too early an age is prejudicial to the safety of the mine, and have given recent instances where so many as thirty deaths at a blow have been caused by a boy of nine years of age."

The Committee went further. The practice of employing infants was, they said, "inhuman." Explosions, while not all causing a loss of life, constantly occurred because children were employed as underground trappers.

And yet "the fault cannot reasonably be imputed to the poor children, ignorant of the consequences of their acts, cooped up alone in the dismal silent darkness of the pit, sitting for twelve hours unrelieved from the monotony of their solitary employment."

The indictment, coming from a body of undisputed authority, caused ripples at Government level and by the time the Committee had published their findings in full there had been repercussion throughout the world. A Commission had been appointed, inspectors had been despatched to the coalfields, and their reports were to bring horrified reaction. The public learned for the first time how generations of boys and girls were being exploited in the darkness of the mines. Authenticated case histories were given of children in the Northern coalfield, beginning work as early as four and a half years of age. In Scotland, colliers – in spite of legislation passed forty years previously – worked in conditions of serfdom, regarded by coal-owners as property to be bartered. They heard of girls and women employed as miners. "No horse in an over-loaded coach, no donkey in a costermonger's barrow, few slaves the property of a West Indian planter, experienced the treatment which was the lot of many children in the mines," wrote Walpole.

The agitation which followed the revelations of the South Shields Committee and the later Commission came as a surprise to the North's coal-owners, accustomed as they were to absolute authority and protection from outside influences.

And yet, the primary object of the South Shields Committee had not been to expose the evils of child employment. Their pronouncements on the subject were simply a by-product of evidence heard on the scientific methods of preventing mine explosions. Thus it was that George Stephenson, the railway pioneer, who was himself a product of the coalfield and had begun work at eight years of age, was surprised to find other conclusions drawn from evidence he gave to the Committee.

In a letter expounding his theories on mine ventilation, he wrote: "It is a difficult task to improve the mode of ventilation. There is always a great number of (ventilation) doors required in a large mine, and any one of these doors being left open by a boy falling asleep may cause some part of the mine to become charged with hydrogen gas, and an explosion from that district may follow. . . ."

He had obviously not anticipated the Committee's response in producing his comment as "a proof of the necessity of not employing boys for that purpose, of six, seven, eight, nine and ten years, as too commonly done."

Evidence to show that the employment of infants was not only inhuman, but positively dangerous to others, was also produced in an analysis of all the *known* explosions which had taken place in the coalfield since the earliest records in May, 1658. (The statistics were not necessarily accurate because, prior to 1814, no inquests had been held on miners killed underground. Fatalities were considered inevitable.) They showed that an explosion at Jarrow Colliery on 17th

January, 1826, which claimed thirty-four lives, had been caused by an eight year old boy called Norman, who had left open a trap-door. Another explosion, under similar circumstances at the Bigge Pit, Willington, had been caused by a nine year old who had left his station to play with two friends. Thirty-two lives were lost, including four boys under nine years of age.

Boy labour therefore stood condemned by the South Shields Committee, but their disapproval would have been insufficient in itself to have motivated reform. It was not, however, evidence produced in isolation; it coincided with a report on the education of miners' children in Wales, made in 1840 by Seymour Tremenheere, a Government official described as a "departmental jack of all trades." There was further evidence heard from an inspector appointed under the 1833 Factories Act who had, incidental to his duties, undertaken an inquiry into conditions in the mining area to which he was assigned.

And, had it been heeded at the time, there was further evidence to be considered following the report of a Government select committee. This was published in 1835 and, in the course of considering how best to deal with accidents in the mines, the subject of infant labour was frequently referred to, but subsequently ignored.

Among witnesses heard was Anthony Winship, a wasteman employed in Lord Durham's mines, who was asked how many children were employed. He replied: "A great quantity; in our concern perhaps some hundreds."

He was further interrogated. "How long do their children work?" he was asked. "They stop twelve hours."

They work by the day? – "Yes, part of them work by the day; but some by what they can get by working piece-rate."

Then they work longer than the men on average? – "Yes."

Do they go down as early as ten years old? – "Yes."

They are generally employed in light work? – "Yes, such as opening doors and getting horses, and so on."

Do they attend any school? – "When they have a holiday."

Is it common for them to take any food with them down into the mine? – "There are no regular hours for meal time. They catch it when they can," said Winship.

Evidence was also heard from Nicholas Wood, one of the North's most pre-eminent colliery "viewers," or managers. He was in charge of Lord Ravensworth's collieries, and had additional interests in other mines at Gateshead Park and Tyne Main.

Asked if he thought it justifiable to employ children as trappers in mines prone to explosion, he was explicit. "I should think that the expense of employing men in such operations would render a great many collieries unworkable to profit," he said.

Would men be more attentive than boys in that duty? – "I have some doubts upon that. I think the boys are generally very attentive. In the case of Killingworth, where I believe there are about fifty boys employed, we have very seldom had to complain of neglect."

George Stephenson, the same who later gave evidence to the South Shields Committee, was asked about the ages of children he personally knew in the mines. Were they as young as eight or nine? "I believe younger than that. I believe children of seven or eight years old have been put in those situations (as trappers). Sometimes it has happened that the father has carried the child to the colliery on his back, he being so young. But that, I think, is not the case now."

Stephenson's evidence was subsequently proved wrong on two counts. Children were at that moment working below ground who were only five years of age. And they were still being carried to work on parental shoulders, as the Commission was to prove seven years later.

"Do you think the health of children suffers," Stephenson was asked. – "I am not aware that it does. I think the pitmen generally are healthy men. They are very cleanly men in the North of England; more so than any other part of England I have seen," he replied.

There was evidence, too, from John Buddle, a man generally deemed pre-eminent in mining circles, whose contributions to the development of the industry were to bring him international fame. Buddle, born in the North Durham mining hamlet of New Kyo, was in charge of Wallsend Colliery and his presence before the select committee was distinctly apt to its purpose – on 18th June, just thirteen days previous to his appearance, Wallsend Colliery had exploded with the loss of 102 lives. Nine of the victims were under twelve years of age.

Buddle's evidence gave the Committee indication of the sort of dangers boy colliers were being exposed to from the moment they left the surface. He recalled that six weeks previously, five men had been killed at Whitley Colliery when a "basket" taking them below ground slipped off its hook and plunged them down the shaft. Boys, he said, rarely sat in these baskets, preferring to cling to the rope itself.

"The boys always do so; they are trained to it. It is what we call 'topping' " he said. "I approve of this method. It trains them to a degree of activity, vigilance and care, which they would not otherwise acquire."

Had no accidents occurred because of this practice? – "I have never almost known an accident arise from the 'topping.' I have known a boy to fall asleep with his arms clasped round the rope, and gone to the bottom of a 100 fathom pit, and the banksman took him off again when he came back. He was not aware of it 'till then awakened by a slap. They don't show any repugnance to go down. It is fine fun for them."

It took evidence from a doctor to put a different aspect on the employment of boys. Dr. George Fife, of Newcastle, a former colliery doctor, was asked if a child of ten was fit to perform the type of duties required of them in coal-mining. "I should not like to trust them if I was in one of the workings. The minimum age I would suggest should be twelve to thirteen," he said.

The Committee, in spite of the evidence furnished, failed in their conclusions to draw public attention to the problems and dangers attendant upon the employment of infants. In fact, they found it "impossible to lay down any precise directions, or to form any rules of universal application." They thought, however, that benefit could accrue from visits to the coalfields by some of the country's leading scientists. Indeed, they reported, they had been assured that such visits would be received with pleasure by the mine-owners.

It takes the benefit of hindsight to fully appreciate the Committee's reluctance to recommend any kind of legislation. Among its appointees were such illustrious names as Pease, Ord, Ridley, Lambton, Bowes, Lowther and Durham, all of them coalowners.

Thus, it was not an official Government inquiry which first threw light upon a dark side of social history; it was the work of philanthropically motivated individuals of whom those concerned in the South Shields Committee were foremost. Their conclusions were noted by Lord Shaftesbury and his supporters. They took it up in the House of Commons and the Royal Commission which followed came at their urging.

Truth, in all its starkness, was about to be revealed. . . .

CHAPTER THREE

"*Come, stranger, come, and see the pit,*
It cannot do you any harm
To see how we, by work and wit,
Through all the winter keep you warm."

— Descent of a Coal Pit.

John Roby Leifchild, an inspector appointed under the terms of the Royal Commission to enquire into child employment in the North East coalfield, faced a difficulty he could not have envisaged when he stepped aboard the North-bound coach from London. His arrival in Newcastle, then in the throes of transformation under the direction of entrepreneur builder Mr. Richard Grainger, had left him bemused and facing a basic problem: how did one communicate with the provincials?

His reception in the colliery villages proved distinctly chilly. Not only did the collier folk speak a dialect so unique it might well have been a foreign language, they manifested distrust and doubts about the purpose of his mission.

It was January, 1841, seasonably bleak (the Rivers Tyne and Wear froze over), and Leifchild decided that his first duty was to learn the language of the Northumberland and Durham pitfolk. He had thought of hiring an interpreter but, after consideration, felt this would defeat the object of his purpose. Irrespective of the difficulty, he must learn to converse in the regional patois. Throughout January and February, he therefore applied himself assiduously to the task. He learned that a "singing hinnie" was an item of staple diet, to "cowp" was to tumble, a "double neef" meant clenched fists, a "backskin" was a leather back protector worn down the mine. His vocabulary increased to the point of understanding that when a trapper boy said his "head was warking" he had a headache. A "thirst," he discovered, was not necessarily to be sated – it meant a worked-out part of a mine. A "thill" was the mine floor, a "switchin" was a colliery siding, a "wailer" was a boy employed on screens.

Hitches, hoggers, crook yer hough, crosscuts, kibbles, nueks, mushy styen, baits . . . they were indeed strange and foreign-sounding words, but each was functional in its context and Leifchild, with difficulty, absorbed them.

By March, he obviously felt sufficiently secure in the dialect to embark on a tour of the coalfield, and throughout the next two months he began to see a new and different England to the one he left behind in London. It was an England peopled by sallow faced men, with a deep-rooted suspicion of anyone they deemed "superior." Men, he noted, kept to themselves and masters sat with masters. Conversations between them were rare. "It would seem to him to be a truth, established by experience, that his master can have no desire to benefit him, and expression of such a wish would disguise unpalatable propositions," Leifchild noted. Meanwhile, the masters held to an opinion that to allow the men any property was "little short of encouraging them to store up an armoury of weapons ready to be turned against their masters. It would supply them with sword, shield and armoury for conducting the war. . . ."

Not only did these peculiar Northerners speak a different language – they also had distinct physical characteristics. They were small of stature and walked with a body curvature, the result of working in low places. They were inclined to bandiness, too. The children, small, bony-faced, pale, were beginning to show the effects of heredity. "Here is a diminished race," reflected Leifchild sadly.

His passage through the colliery villages was met with scarcely veiled distrust among those whom it was his task to interrogate and it quickly came to him that the resistance was not entirely natural, nor simply due to his presence among them as an inquiring stranger. His approach to youngsters was met with silence or downright rudeness. Some parents, on seeing him approach, grabbed their children and hid them. When he finally did manage to engage one in conversation, others came to shuffle the child away.

It was some time before he came to realise that the opposition to his presence was as much the doing of employers as of the pitfolk. He was aware, of course, of the opposition manifested towards the Commission by the great colliery owners, but to meet it in practice, in the mean colliery rows, brought home to him the power wielded by those coal barons. Freedom of expression, in the coal districts, came only by permission of the masters, it seemed.

Leifchild persevered and, as time progressed, he was to become completely fascinated and enthralled by the sights, sounds and the people of the area allocated to him by the Commission. He had arrived at a time of burgeoning progress in the coalfield. What emerged from his investigations was therefore not a mere parroting of evidence heard on the practice of employing children underground, but a report which gave a word picture of a way of life in the throes of upheaval and development. He was to give a new view of a region of England which, by geographical isolation alone, had developed distinct and unique characteristics.

He moved into the outlying areas and saw pit villages in the making. First would come the shaft sinkers, followed by the jerry builders who erected the long rows of one-up, one-down cottages, to house the incoming miners. Then the colliers and their families would move in on wagons piled high with furniture. Some would be established colliers, moving to the new village for the sake of change or, perhaps, slightly better wages. They would form the professional cadre of the new community. The balance, in Leifchild's eyes, appeared as "the very scum and offscourings of a peculiar, mischievous and unlettered race."

The North, at the time, was in its Klondyke period of development with new pits being sunk and new areas of development being explored. Leifchild's "scum and scourings" were more likely to consist of poor Irish and Scots falling to the lure of "high" wages to leave forever the rural areas of their birth. Once claimed by the collieries, few were to escape them. A system of "binding," plus credit at a colliery-owned store (a system condemned and banned a decade previously, but which still flourished well into the mid-Victorian years), ensured a constant labour force for the pits. Among them, too, would be tin-miners from Cornwall, swarthy, brown-faced men, bringing with them their own language and traditions of mining. Welshmen, chasing dreams beyond the harshness of their lives in the mining valleys, came by the score to find their hopes crumble at collieries where conditions were a replica of their homeland.

This, then, was the "Other England" now being explored by Leifchild and, at every turn, there was fascination for him. He noted the lack of sanitation in the new villages, with the space between two colliery rows occupied by an ash heap and dung hill, among which the grubby toddlers played their games of marbles and mock pitch-and-toss. There were no conveniences, and baking facilities consisted of one communal oven per row. At some doors stood a ramshackle pig-stye, with a snorting occupant awaiting the Christmas kill. And then, in odd contrast to the obvious, outward squalor, came a view of the interior of the cottages, each with an eight-day clock, a mahogany chest of drawers, and, in splendid occupancy of the upper room, the sort of four-poster bed which could well have graced the home of the rich. There were, in some cases, desk beds for the children and even when the bedding was laid upon the attic floor, it was generally clean and of good quality.

He noted, too, the pattern of daily life in the villages, with the constant surge and retreat of men at the behest of the pit. He saw them as a tribe of "blackamoors" returning home at evening with the dust of the pit on them; he observed again the physical peculiarities of the native miner, with his bow-legs, the curve of the body, the protruding chest, and the eyes, hawk-like. Each moved to his home for a ritual scrubbing, and then, almost as if on signal, through the half doors and into the street outside cascaded the soap suds left by their ablutions.

Evening, in the villages, saw the inhabitants engaged in slow promenades and quiet conversation, with pipes emitting clouds of pungent smoke to the background scraping of a fiddle or lilt of a flute. The Methodists went to their tea meetings, the drinkers to the alehouse, and the sportsmen to the various contests arranged by themselves.

Leifchild very quickly noticed, however, a strangeness about these village scenes – where were the children? Babes in arms were plentiful enough – large families were common-place and it must have seemed that every woman still of child-bearing age had a child nursed to her bosom, but it was the scarcity of children from the age of five upwards which was noticeable.

A full four hours – perhaps even eight – would elapse between the return of men from the colliery and the appearance of the missing age group. Then, at a given hour, they would be seen trooping home from the colliery. It was a sight guaranteed to stir the hearts of the uninitiated, and Leifchild himself was not immune. Children, he learned, habitually worked twelve hour shifts and more, and would remain in the mine long after the older men had returned home.

Coals were drawn from the collieries twelve hours a day, on average, from 4 a.m. to 4 p.m. The coal hewers worked eight hour shifts, but the boys were below as long as the pit continued to work. At the very minimum, it meant they were away from home for thirteen hours.

And even those hours were optimistically assessed. In the months of investigation to come, Leifchild was to learn of children who habitually remained below ground for periods of twenty-four, thirty-six or forty-eight hours consecutively. He came across the case of one boy in the South Durham area who had remained down the pit for an entire week. Those were the conditions endured by the youngest boys. He heard of others, admittedly older, engaged in the strenuous work of "putting," who would work until only physical exhaustion prevented them carrying on.

"Children – trappers – would be deemed by some to be the objects of unmitigated misery," he was to write. "That conditions were mentally and physically pernicious could not be denied. . . . It is no unwarrantable assumption to affirm that the unbroken monotony of (their) duty in conjunction with its duration and the darkness, solitude and other peculiarities . . . must at least blunt the feelings and deaden the intellect so as to diminish the capability of receiving instruction. The perpetual recurrence of the vacant stare and of the heedless "I don't know" in answer to reiterated and varied questions . . . were of themselves sufficient symptom of this morbid condition."

Leifchild had his troubles – the communication difficulty caused him genuine concern until he was able to gain some semblance of the dialect himself – and critics were to decry his methods of investigation.

One oft-repeated slur was that he spent a too high proportion of his time in the homes of the coal masters and was inclined to accept their assurances as evidence. That he did so to a certain degree is indisputable – his reports give indication of it, as witnessed by the number of interviews conducted with colliery viewers and minor officials. He published, for instance, the comment of one leading official, who referred in scathing terms to any evidence gathered from the children in his employ. "Their habitual vice of lying and deceit is so inveterate that no information from them should be trusted," he said.

Leifchild himself could be scathing in his references to the older miners (who were, in any case, outside the scope of his inquiries). "For the stranger to read the mind of a pitman, a circuitous approach and no small tact is required," he said. "A prominent feature in his character is jealousy of superiors and deep rooted suspicion of employers. . . ."

"Conversation at length being practicable, the ignorance, obstinacy, doubts and suspicions of the witnesses were developed. They were too combated in nearly every individual and some were more disposed to put questions to me than answer those which I proposed to them. Several young persons absolutely refused to make any statements whatsoever."

They were comments which made him ripe for criticism from the social reformers who had urged the need of a Commission – but he could be equally critical of the masters. He felt, for instance, that the object of his mission was "misunderstood by the coal-owners and misrepresented by their hirelings." He was also critical of the type of house supplied by the coalowners for use by employees. He thought conditions "brutalised" the pit folk and condemned the system in some areas, whereby speculators were allowed to build the colliery rows to let them off to the coalowners at £3 to £4 per annum.

It was, he said, undeniable that the masters could materially add to the physical comforts of the pitmen by providing baths (he was before his time) and village drainage.

He also faithfully recorded some of the outlandish comments made to him by the more retrenched masters. One viewer, for example, a man in charge of three of the biggest collieries, asked Leifchild to observe that the Northern colliers were "a healthy race – you have seen that their cottages are very comfortable?" The man prided himself on introducing horses to his colliery to dispense with certain operations of "slavishness" but then confessed that his motive was economic as opposed to humanitarian.

"But then," he added, "the result is the same. And it has the advantage of being much more certain than the doubtful operation of man's humanity to man. . . ."

The same viewer must have been surprised to find one of his collieries earned adverse comment from the commissioner in his report.

If the proofs of the need for reform were needed, Leifchild was to have them quickly thrust upon him. Within the first three months of his investigation, at least four mine accidents took place. On 5th January two men were trapped in Elvet Colliery, Durham, by an inrush of water. On 14th January the ill-fated St. Hilda Colliery, at South Shields, caught fire and almost claimed the lives of twenty-two men and boys who were below ground. On 15th March four men died in a shaft accident at Cowpen Colliery. Leifchild made no mention of any of these in his final report – but on 19th April there occurred a major disaster which did claim his attention. At 1 p.m. that day, the pit at Willington, property of Matthew Bell, M.P., and Partners, blew up majestically and claimed the lives of thirty-two of the thirty-five men and boys below ground.

The cause, it transpired, was the result of a trap door being propped open, while the infant in charge left to play with two friends nearby.

Leifchild made his way to Willington. . . .

CHAPTER FOUR

"*Why must I sit behind the door*
So many hours away from you,
And hear the putters shout and roar
And naught but shut and open do?"
— The Trapper's Petition.

The scene at Willington Colliery, about a quarter of a mile from Wallsend village, was grimly typical of its kind. The sound of the explosion had been heard throughout the area and crowds of apprehensive relatives were quickly joined by colliers from the surrounding pits who volunteered to go below in rescue parties. The explosion had taken place in the "Bigge Pit," one of three shafts which had only recently resumed working. The prize, for the owners, was the Bensham seam, a four foot band of the finest coal the Tyne coalfield could offer. It was also a notoriously explosive seam. An attempt to win it in 1831 had resulted in an explosion in which three men had been killed outright and fourteen others severely burnt. Four subsequently died. Gas had continued to pour into the seam in such proportions that it had afterwards been bricked over in the manner of a gasometer, and the workings abandoned.

There had been other explosions in other seams at the colliery. In 1829 four men died when an explosion released a flood of water from old workings. In March, 1840, one man died and three men and five boys were severely injured in an explosion.

Now, the pit had again exploded and this time an official Government observer was on the scene to record events. The workings, being new, were not extensive. At no point did they extend beyond 220 yards from the shaft and it was a source of amazement and conjecture among experts at the scene that so much gas could have accumulated so quickly. The three survivors – two men and a boy called Charlton – were hauled to safety badly shocked by their experience, and the melancholy task of raising the dead began immediately. It took three days to complete and, following due deliberation, the experts gave out their verdict on the cause of the explosion.

The fault, indeed, could be attributed to that small trapper who had left open his door to play with colleagues. This had disturbed the system of ventilation and allowed an explosive accumulation of gas.

The verdict was accepted with resignation by the sorrowing onlookers. There was little remonstrance. Such accidents were regarded as inevitable in a coalfield almost innured to the sights and sounds of death in the mass. The fact that thirty-two men's lives had been entrusted to an infant who had failed in his duty struck no-one as odd. These things happened. It was the way of things.

The colliery viewer, Mr. George Jobson, evidently a man of strong opinion, was asked about the employment of child labour and remained convinced that it was to "the national good." Jobson himself was a substantial employer of infant labour, both at the Willington Colliery and at others he managed at Heaton and Burdon Main.

In his comments were capsuled the attitudes of the Northern mineowners and the leading officials of the time. Here was laisser-faire with a vengeance.

"The children stand in awe of deputies and superiors better than the older ones. Old men fall asleep and fit men would never stand the degradation of doing a trapper's work," he said.

He thought that the existing Factories Act which gave protection to children could not be held relevant to pit work. "With the present system of working hours, children could not go to school before work and would be incapable of receiving it when the pit closed down at the end of the day," he said. He refused to suggest any system whereby children could receive education and, in fact, actively disapproved of it. "Over-education might make a man unfit for labour. I don't promote men because of education."

How, then, did he select men for promotion? "I have promoted one man from miner to underviewer – but that was because he was a steady and efficient workman of sound common sense, not because he was educated. The man was ignorant before, and only picked up the rudiments after promotion," said Mr. Jobson.

He thought that perhaps three-quarters of his boy employees would receive some rudimentary education at the once-weekly Sunday School held in the village. A few of the older men had taken up the study of "natural philosophies" but Jobson remained unconvinced of its worth. "Too much time spent in such studies will make them dissatisfactory workmen and won't qualify them for superior positions. In ordinary matters, education makes no difference to them as workmen – conversely, in strikes, it is the educated ones who are always the hardest to deal with. It is not always that they take the correct view of the nature of the disputes – it is more from self-sufficiency and self-satisfaction with their superiority."

Among the worst such offenders were, he thought, the Methodists among his employees. They invariably acted as spokesmen during disputes.

Education was therefore unnecessary and an actual hindrance to the profitable working of coal mines. The only thing he could deem desirable was any acquaintance among the men with the nature and mode of working coal and, with coal hewers, this was actually both desirable and beneficial. He felt that the men themselves would oppose any move to establish schools in the villages. In any case, the owners of coal did not work in it and largely left the extraction of it to lessees whose interests were to make the greatest profit in the shortest possible period of time. The lessees had no permanent interest in either the mines or the people employed in them. The workers invariably changed from year to year. Schools could not be conveniently sited; they would be sparsely attended and pitmen, generally, were just not interested in education. In his experience, he had known only one miner who took an interest in general literature.

Meanwhile, he strongly contended that the child miners in his employ were healthily engaged in a worthwhile pursuit. "Their conditions are good," he asserted. "Occasionally, a boy might be down the mine all night for some extraordinary reason such as cleaning horseways, and he will be followed by other boys at 2 a.m. Usually, they are down for twelve hours when a signal is given and a deputy or a man will go round to see that all lights are out before leaving the pit. Twenty-five years ago, the boys often worked double shifts without coming to the surface at all – but they were paid high wages. The position of a trapper is very secure and strong timber supports the roof where he sits. A large place will be hewed into the wall for him to sit in, and he need never leave the place, having a cord to open and shut the trapdoor. A light is often – or very generally – placed near him."

At what age did he take on his boys? He said that he had known boys employed at six years of age, but there were none currently employed in his mines below seven. He thought they were now treated kindly by their parents and considerately by the coal owners – but even then, when they got to eighteen, they were apt to become most ungrateful.

"Personally, I do not think children should go below before they are nine," he said ". . . but this should not be subject to law or legislation." Conversely, if the age of entry was later than ten, they would never make pitmen (although there had been cases to the contrary). In all, he was positive that parents would rather see children employed than educated.

Their health, he concluded, was good – in fact, unhealthy children could not be employed in the mine, so again there was no need of legislation for the mines similar to that of the Factories Act. It was, in his experience, unnecessary to employ men in poor health as there were healthy men in abundance. The health of individuals was quite easily ascertained – their work performance could be checked against the record of their first day at work.

Miners, he thought, proved their native, rude health by an ability to recover more quickly from accidents than other classes of work-people.

Finally, he could see no alternative to the employment of children.

"If we employed men at the work, it would bring ruination on the trade. There are not enough old men around to do the work and mechanical means of doing it cannot be found. It is impossible to employ the boys for less than twelve hours without the greatest loss and inconvenience to the coal-owners. Any legislation on these points would be quite unnecessary and prejudicial to the owners and the workmen. It would induce habits of looseness and carelessness," concluded Mr. Jobson.

The roseate picture of healthy children fulfilling a useful and well-rewarded function essential to the economic health of the coal trade contrasted sharply with that painted in conversations with the children who were employed at Willington.

Leifchild himself went below to talk to them but found it an impossible task. He found the "lads" in a state of perpetual motion. He had to wait until their arrival on the surface.

He had gone to the pit immediately news of the explosion arrived and was shocked by the sight of the dead being raised to the surface. He found particularly poignant, the maimed bodies of the four children involved. The death roll included two ten year olds, one aged eleven, and one thirteen, all doorkeepers.

When the pit resumed work again, among those to go below to his duties by the trap door was a diminutive boy who gave his age as six. He had already worked below ground for six months and was following a family tradition of mining. His father was a hewer and his two brothers worked as a putter and driver respectively. He arose daily at 3 a.m. and walked to the pit to start his duties at 4 a.m.

The pattern of his life was pathetically echoed throughout the coalfield. "John Black and the other rolley drivers sometimes beat me with their hands, but I am not frightened now. They use their whips a little, but not to hurt. I once went to Sunday School, but not now," he said.

Another boy, William Ransom, did not know his own age, but was thought to be around seven. He had already twelve months underground experience. He sometimes worked late; how often he was unable to state, but his dad was a colliery deputy. He, at least, was not beaten. He worked a quarter of a mile from the shaft bottom.

A total of twenty-six pony drivers were employed at Willington. They were aged between ten and thirteen, and had been promoted to the position following service as trapper boys. The record of injuries they had suffered in the mine read like a roll call after battle. William Cowell told how a stone had fallen on his head and put him out of work for three months, and pointed to scars on his head as proof. A boy called Crosby had been kicked by a horse. Another had injured his foot when a rolley passed over it. He had been off work for two weeks but had returned to work before it was fully healed. Three others had been hurt by rollies and one had broken a finger.

Mr. Jobson's assurances that his boys rarely worked beyond the usual twelve hours did not accord with the accounts given by the boys themselves. John Nichols, an eleven year old, had worked two twelve hour shifts without break, returned home to sleep, and was back for a third shift twelve hours later – a total of thirty-six hours out of forty-eight. Almost all could recall having worked double shifts.

Time, for the majority of such children, ceased to have meaning from the moment they climbed onto the rope to descend the shaft. Work became all-consuming, and they were left with nothing outside the pit by which to judge chronology. The passage of days and months could only be assessed by the number of accidents in which they were involved. "I don't know when it was, but it was just after my last 'laming'," said one. Another said that he had learned to read "the last laming but one." Some judged their age by their transit from trapping, to driving, putting and finally hewing. A fourteen year old recalled that he was lamed three times as a driver, and once as a half-marrow.

In subsequent investigations Leifchild was to hear of a boy who had lost a finger at Felling Colliery, broken his leg at Walker, and cut his face at Percy Main. Another was lamed twice at Howden, but only once at Walbottle.

"If they are unacquainted with learning they are well versed in laming," he commented wryly.

In one pit alone – East Holywell – where fifty boys were employed, it was recorded that eleven alone had lost 154 working days between them, through falls of roof and tramway accidents. They were jammed between trams, had ribs broken, legs maimed and hands damaged.

And all for wages ranging from sixpence to ten pence a day.

CHAPTER FIVE

"I'm up in the morning afore it is light
And down in the pit in the dark;
And tho' I get home afore it is night,
I'm asleep from my terrible wark."

In the year 1841 there were forty-six collieries shipping coal from the River Tyne. They employed underground, by the best estimate available (considering the not unnatural predilection of the coal-owners to obscure the truth) approaching 1,500 children under the age of thirteen. Between thirteen and eighteen, the figure was assessed at 2,000. The adult work force totalled 7,261. In the South Durham portion of the coalfield, approaching 200 in every 1,000 employees was under thirteen. Wearside collieries employed 9,000, with similar proportions of infant labour.

It appeared to be a general rule that the smaller the colliery, the higher the proportion of boys employed. Percy Main Colliery, for instance, had a total work force of 250, of whom 105 were boys. Washington Colliery had twenty boys out of ninety employees. At Cowpen Colliery, one-seventh of the employees were boys, from a total of 151. Seghill had fifty-two boys, Burdon Main thirty-six, Waldridge twenty-three, Coxlodge forty-four, and Wallsend thirty-three out of a total of 112. At Hartley Colliery – a particularly accident-prone pit – there were 140 employees of whom forty-four were between thirteen and eighteen and twenty-eight under thirteen.

In the large, newer collieries, where advantage was taken of the advancing technology of mining, the proportion of boys employed was smaller.

Such was the case at Monkwearmouth Colliery, the deepest pit then sunk, and proudly deemed by its owners, Messrs. Pemberton and Smith, to be one of Britain's wonders. The shaft was four times as deep as St. Paul's was high, in total 1,680 feet from surface to water sump. Descent was by a "tub," held by a rope weighing five tons, which could convey coal to the surface in thirty hundredweight loads. It took up to three minutes to reach the working seam – the Bensham – and so rapid was the descent it made the ears pop.

The colliery was managed by Mr. George Elliott, aged twenty-seven, of Penshaw, who was held to be an enlightened employer. He had even applied his ingenuity to the problem of dispensing with the need of infant trappers and had recently introduced a spring loaded trapdoor which many thought was an efficient alternative. Elliott was the first colliery viewer to try the new mechanism, but only on an experimental basis.

His pronouncements on the problems of child labour could also be deemed radical in the context of comments made by other viewers. Elliott thought that twelve was soon enough for any lad to commence work in the mines, although owners Pemberton and Smith felt that ten was better. He also refused to join other viewers in their condemnation of education for children and had even gained the support of his employers in helping "the Dissenters" to build a school. Nor did he dislike those Methodists in the mines who, by self help, had educated themselves into moral leadership of their colleagues. "Methodists have a great hold on pit people and, although I am not of that persuasion, I must say that Methodists have done more to ameliorate the condition of the pitmen than the whole Church put together," he said. "Those of my employees who are less morally stable owe it simply to the sinister influence of the town of Sunderland just across the water," he added.

An enlightened employer . . . and yet, he himself confessed that he had recently employed a boy named Robert Pattison whose age, he later discovered, was only six, and if his attitude to his infant miners was one of basic sympathy it made little difference to them once they got below to their working stations.

Alexander Bell, a pony driver, told how he was struck by older men when his wagon – through no fault of his own – jumped the rails. Sometimes they flung stones at him, and his overman – a man called Robert Robinson – often flogged him with a stick "as hard as he can." This had just happened, he said, and once he had been so severely flogged he was too ill to return to work.

And yet, the Monkwearmouth Colliery was among the few where visitors could expect a welcome and perhaps an invitation to view the mine in operation. The sheer depth of the workings made it a source of wonder to those outside the industry. The shaft had been sunk against professional advice, and at great expense, in a part of the coalfield which had hitherto remained untapped.

The open-handed attitude of Elliott and the owners may have been as much an expression of pride in their achievement as a desire to publicise the industry, but visitors did come to express their wonderment and, on occasion, give amusement to colliers long innured to underground terrors. Those terrors were apt to be more real in some of the older collieries. There, strangers were not tolerated to the point of violence. In times of strife no senior official would dare venture below and men sought by authority could be spirited below and remain safe from all pursuit.

It therefore came as a surprise to "outsiders" who ventured into Monkwearmouth Colliery to find the employment of infants so blithely tolerated and the sight of them was to trouble their conscience.

Leifchild himself described in later years the effect upon him of his own visit to Monkwearmouth. He met in the pit one boy, aged seven, whom he thought was at least a year younger. The boy had been down the colliery eighteen months previously when his father had been killed in an explosion and the memory of it still troubled him. The same explosion had killed a colleague trapper named Cooper and the boy recalled that the blast had wafted over him "like a heavy wind and it blew all the candles out."

"This – in Merrie England!" mused Leifchild.

Visitors, not yet weaned on Dickensian pathos, found the sight of the trapper boys in their lonely stations underground to be an aesthetic offence and an assault on finer feelings. Usually, boys were supplied with two candles by their parents. Ideally, they could be made to last through the shift, but if family finances began to fail towards the fortnightly pay-day, then the trappers' candles were traditionally foregone.

Thus, the visitor stumbling through the galleries would hear a small voice emanate ethereally from some dark and unexpected corner, with a plea to "Give us a low (light) please, mister?" Alternatively, the request was "What o'clock is it?" or "Will it soon be time for kenner?" Kenner, they learned, was the word which lightened all pit activity and gave motivation to the working day. "Kenner" was the word shouted down the shaft by the banksman to bring each shift to a close.

Loneliness and lack of illumination were not the only problems faced by the trappers at Monkwearmouth. They, in common with other employees, faced a health hazard, which was unique to the colliery and led, at least in the early months of employment, to curious outbreaks of sores on their skin. It was thought to be due to the depth of the colliery and the subsequent increase in temperatures, eighty-nine degrees in some parts, together with the chemical properties of the water through which they splashed underground.

A boy called McKenzie told how his first three months at the pit afflicted him with ninety "boils" on his back which erupted in two's and three's and lasted about a fortnight. The water was both salt and hot, and had brought out yet more "boils" on his thighs and arms about the size of an egg. All men at the pit would suffer at some time, and a remedy which received favour among them was a poultice of turpentine and linseed.

The heat was such that all except drivers and trappers worked naked, apart from a flannel loincloth and shoes. Water was drunk copiously to replace lost perspiration. A young putter, Alexander Ball, told how the sudden transition from the heat of the pit to the cold of the surface air repeatedly left him with colds. For George Allen, a driver, the worst complaint was that water, dripping from the roof, would "hurt his boils."

Dr. W. J. Todd, employed as the colliery doctor, investigated the phenomenon and concluded that the boils, or carbuncles, were simply the hallmark of a man or boy fresh to the pit. "They are affected first by the boils which are brought on in the first instance by the heat, and then aggravated by the salt water. This water pours out in quantities from dykes and troubles (geological faults) which are frequently met with in the earlier working of the pit. The water is not caustic, but it is extremely irritating. After a time, perhaps two or three months, the men become wholly free from the boils and remain so ever after. It is more like a carbuncle than a boil and discharges a little. An aperient medicine and poultice generally suffice, but they are very troublesome and irritating for the time they last." He thought that the coal hewers suffered more from the condition than the boys, and the occurrence was confined to new employees. The water of the pit was likened to the diluted brine of a pickle tub.

Dr. Todd also thought that the stature of the children employed at the showpiece pit was "a little below that of other labourers."

The phenomenon was exclusive to Monkwearmouth and was unknown at a neighbouring colliery at Hetton, five and a half miles east of Durham city. This immense colliery, and the attendant village, was also exploiting the hitherto untouched coal seams which lay towards the Durham coast. It had been sunk, again at tremendous expense, in open country in the vale of Houghton.

Until the shaft sinkers moved in it had been desolate and barren, but by 1840 an entire township with a population of 2,000 had sprung up around the pithead stocks. It was boasted that sufficient coal was raised in an eighteen-hour period to fill thirty Thames barges – about 600 tons. In 1822 an eight mile wagonway had been constructed to the staithes on the River Wear. Conveyance of the trucks was powered by "five of Mr. Stephenson's travelling engines" plus two fixed engines, each of 60 h.p.

It was an immense undertaking, but success had come with a price. The miners, drawn from all parts of the country, had proved troublesome and in times of strife had revealed a distinct predilection for violence. In the great strike of 1832 a blackleg pitman named Errington was shot dead in his home. The population had gathered to jeer as his coffin passed along the streets. Retaliation had come in the form of evictions and detachments of the new London police, and the Queen's Bays, were posted around the town to quell any disturbance. The miners themselves were armed, and shots were exchanged by both sides before the evictions could take place. The new town of Hetton did, however, achieve fame in other directions and is still remembered today as the scene of the earliest attempt at Co-operativism in the North East. Its prophet was a miner named Mackintosh, but his initial success was frowned upon by the masters and they crushed it out of existence. Mackintosh was forced to flee to America and never again returned to England.

By 1840 Hetton had settled into a steady, if not quite sober, community, noted for the quality of coals produced, and as the centre for some of the North East's best dog-fighting. (It was also a noted home of cock-fighting, but the police had only recently cracked down on that sport.)

The colliery had a workforce of 528. A total of 179 were youngsters employed both above and below ground, and the pit claimed its victims among them. At every trapdoor sat diminutive, dusty-faced infants, plying their lonely trade. The driver lads sweated and urged their ponies to yet greater efforts; the marrows, half-marrows and putters, heaved and strained at heavy tubs. Among them were the three Lowry boys, the oldest of whom, at seventeen, already had nine years experience of pitwork. Back home in Downs Lane sat Mrs. Lowry, who must occasionally have pondered on the price one family had to pay in order to exist. She had already lost two of her other boys, one by a fall of stone, the other when he fell down the shaft. There was little sympathy from officialdom. "The Lowry's are generally known to be a dissatisfied family," said an underviewer.

CHAPTER SIX

"*Ye pitman lads, so blithe and gay,*
Who meet to tipple each pay-day,
Down on your marrow-bones and pray,
Success unto the Coal Trade!" — The Coal Trade.

The bleakness of life in the mining villages of the 1840's, with the pervading atmosphere of degradation, was apparent to all outsiders who came to view them, and yet the outward visage of the colliery rows cloaked the redeeming facets.

There was a very real sense of identity among pitmen, who were apt to regard themselves as an elite corps of workmen (and, indeed, were) and adopt a certain swagger not noticeable with factory operatives.

Why, then, did parents permit, or at least submit, to the employment of their children underground? Were they all like the Monkwearmouth miner, who was alleged to have begged the master to hire his six year old son? Were the miners so impecunious that they could scarcely wait for the cradle to cease rocking before they sent off their children to the pits?

There were such parents – but the truth behind child employment was considerably more complex than the mere manifestation of parental greed.

Children were cheap. Every infant employed meant a man's wages saved. Thus, when jobs were scarce and men scoured the countryside for work, those with the biggest families of male children would be the first to be accepted. There were cases, too, where fathers had been killed in the mines and left behind large families without support. These were the truly unfortunate. Widows had no security of tenure and – without children, or left with only babes-in-arms – the treatment they received was short and brutal. They were evicted.

Should the widow have boys aged from four and a half and upwards, then the master's charity would be forthcoming. He would offer her children work, and allow her to retain the colliery cottage.

Such cases were numerous throughout the Northern coalfield. There were widows a'plenty – a conservative estimate of the number of miners killed in accidents in Northumberland and Durham over the preceding forty year period was 1,500.

The figure was doubted, even at official government level. When mine managers were asked to return particulars of fatal accidents, the manager of the Whitley Colliery, a small concern, stated that no accident attended with serious injury had occurred for two years. Yet even the local papers carried an account of an explosion which had taken place just twenty-five days previously, when an under-viewer was killed. Government officials reported that the incident was regarded of so little account that it had genuinely been forgotten.

Widows, then, were not scarce in the coalfield, and their children trod the same path as the fathers and fore-fathers until creaking age and the poorhouse beckoned. The only thing sure, constant and unchanging, was work. Ralph Hall, an employee at Heaton Colliery, celebrated his seventy-seventh birthday in 1841. He was still working an eight hour shift at the colliery for a wage of 2s. 6d., plus his cottage and a free supply of coals. He had been down the mines for seventy years. "It was worse for trappers when I first went down," he said and there was no exaggeration in his statement. An overman at Wallsend Colliery, with forty-three years service, thought that children were well-off in comparison to those of his youth. Then, they regularly worked seventeen and twenty hour shifts.

In general, however, the employment of children was accepted without remonstrance by fathers who had the benefit of their own experiences to establish the precedents, and it was widely held that to delay a boy's entry into mining beyond ten would bring trouble. The lad would "not settle in."

An illustration of the route by which the average child went to the pits was given with the case of a six year old boy named Thomas Roker, employed as a trapper at the Flatworth.

His mother reported that as a three year old he had been a "fat little boy" but had, since starting work, grown both whiter and thinner. He had attended a school when he was three, and it had been his father's determination to have him better educated before sending him to the pit. He had, however, taken him to see the colliery one day, and the boy had cried until taken to see the colliery viewer who immediately offered him work.

"I always put him to bed early," she said (she was interviewed at 7 p.m. on a Sunday evening, and the boy was already fast asleep). "He must get up every working morning at three o'clock. He will rub his eyes and say he has only just got to sleep. He gets to the pit at four o'clock and gets his dinner when he gets home at half past four, or quarter to five. He will then wash himself and go to bed between six and seven o'clock. He never has more than two hours away from the pit for eating, washing and playing, but when he gets hardened to the

pit his father will stop an hour off his sleep and send him to a night school. Thomas generally goes down the pit in a corf (a coal basket, later replaced by tubs) with a good few other boys in it, and sometimes he goes on his father's knee. It is a dusty pit, but he never complains. The pit does not hurt him, but he tells many a queer story about it. Johnny Giddis was younger than Thomas when he first went down. I think that people send their children down the pits younger than they used to."

She had not heard that the pits damaged the eyesight of boys but she recalled that when her husband had joined the miners' union he had been "sent away." He had finally obtained work at Walker Colliery "where the foul air blinded him somtimes."

At the same colliery was a boy of nine named Pearson. He was even smaller than the diminutive Thomas Roker. His father had been killed in the pit, leaving behind a family of six.

The colliery doctor at Flatworth, Dr. G. W. Huntley, was asked about the general condition of the pit children. "The physical condition of boys employed in this mine is not at all inferior to other children in the neighbourhood. They live remarkably well on four meals a day. Coalminers usually live to a great age. . . ."

Old established mining and industrial areas were particularly prone to the rule that son must follow father to the pit and this was particularly evidenced in the Blaydon and Winlaton area, which was a hotbed of the reform movement. There, even the most ardent reformists accepted child employment, while willing to risk life and limb in the pursuit of suffrage. Ann Mills, thirty-five year old wife of a coal hewer at Blaydon Main Colliery, had borne seven children but only three survived. One began work at the age of seven. Another had gone below when six, due to her husband's ill health. "I wouldn't have sent him, but he would not go to school. I do not think their health is affected – but they do learn to swear. Once last week they got a beating off their father for it." She did, however, send them to Sunday School.

A colliery official said that children were not *compelled* to work in his pit. They were asked just once if they were willing to accept the work, and never asked twice. "In 1839 this place was notorious for Chartism. The 'sacred month was begun' (a Chartist call for a general strike) but the magistrates were very active and were assisted by the military. Four were jailed for refusing to work, and fined. The others who rebelled were dealt with heavily. . . ."

Wherever he moved (and he was to visit a total of thirty-one collieries in the Tyne area alone) Leifchild heard two contradicting stories – one from the owners and their officials, who painted a roseate picture of child employment; the other from the children themselves, who, being largely inarticulate, had to rely upon their interrogator for honesty of interpretation.

Mr. William Hunter, the viewer of Walbottle Colliery, was convinced that credit given to pitmen by local shopkeepers was "baneful," thereby intimating his belief in the former system of colliery-owned truck shops. "They shift around to avoid paying off their debts," he said. "I recall the 1840 bindings at Charlaw and Sacriston when 60 out of 120 families moved out of the area."

He added: "Walbottle is a new colliery. The scum of the other collieries are here."

Mr. Stewart, the underviewer at Coxlodge Colliery, owned by the Reverend Brandling and partners, and notable for an early experiment with steam locomotives, was unsympathetic to any move to dispense with child employment. "It was harder in my days – and I went to the pits when I was eight," he said. "Boys were more corrected than they are now. Now, they only get punished when they need it. The overman will give them a few strikes with a whip or stick, but not to hurt them badly. The beating is no more than a gentle correction."

At Earsdon Colliery, coalowner Mr. Thomas Taylor strongly disapproved of any legislative interference in the running of coal mines. "You will observe that our colliers and their families are a healthy race, and you must have seen that their cottages are very comfortable."

His employees included a William Jakes, employed as a foal and he was observed to look very pale and unwell. Jakes said that he often felt sick and ill and his head ached because of the foul air. His job entailed the heaving of coal baskets through low galleries. "My back often gets skinned and is sore and red now from holding back the corf on the inclines," he said. He had twice been off work because of the bad air, and when below was rarely left with time to consume his "bait." He often vomited and, in fact, had been sent home for doing so. Schooling? "I have forgotten all about it," he remarked.

Killingworth Colliery was generally deemed to be among the best and most intelligently managed in the Northern coalfield. The viewer, Mr. Nicholas Wood, was nationally renowned. He agreed that the children in his mine were perhaps a little "ill-formed" and the majority were pale and not at all robust. Trappers began work in his pit at the age of eight. They were contracted by parents, who "mostly wish to bind them to heavier work than they are fit for, merely for the sake of the additional wages."

He felt it unfortunate that the children of hewers and pit families generally were not sufficient in number to supply the demand, and the boys of mechanics and labourers in outlying areas were being sought to fill the gap.

"Pitmen with families are much sought after. As the pits get deeper we will need more boys," he said. He, too, could not see the need for legislation to regulate the employment of children; and on the subject of education, he thought that higher education was an unpalatable proposition. "The work is merely mechanical and there is no superior mode of performing it. It is the opinion of many that they would be worse workpeople if educated – education could produce a dislike to work. Conversely, he felt that the "popular commotions" in the North East were worse where the population was the most ignorant. His own miners were above the common standard of intelligence and were steadier than those at an adjoining colliery. On balance, he therefore considered that an educated pitman was preferable.

Wood did more than consider the matter. He gave active support to those of his employees who, in 1840, approached him with a request for help to build schools in the village. They had formed their own educational society and had plans for a school for infants aged two to five, a day and evening school for those over five, and a third school for girls where they could be taught homecrafts. Wood readily agreed and the three schools were built and officially opened in August, 1840.

More than 400 of his workmen subscribed three pence a week to maintain them. The sarcastic comments of fellow viewers – "pitmen will not send their children to school" – were silenced when it was learned that 528 pupils had been enrolled. The infants' school was a particularly notable acquisition. It was the first ever to be erected in the colliery districts.

Wood, who acted with the benevolent support of coalowners, Lord Ravensworth, the Earl of Wharncliffe and John Bowes, M.P., was philanthropically out-of-tune with his fellow viewers in the region.

A more typical attitude was manifested by Matthias Dunn, the viewer and part-owner of St. Lawrence Main, a colliery on the eastern outskirts of Newcastle. It was a 100 year old mine which had been re-opened in 1831 following a flooding incident. It was among the first collieries to introduce cages in shaft-winding.

Dunn felt that child employment was perfectly consistent with their health and, although he knew of boys starting work before they were six, he did not think legislative interference was necessary. It could be left to the good sense of parents and the masters. He took particular exception to a suggestion that children should be taught to read and write before commencing work. He much preferred the status quo and while he had heard of boys working for sixteen to seventeen hours he did not believe it caused them any permanent injury.

Alas, a different tale was unfolded in the pit itself. Thomas Wigham, described as a "wee thing of a boy" had been employed since he was five years old and had already suffered his first mine injury – a broken thigh. His brother, aged eight, broke an arm when caught by derailed tubs. He had almost been killed and had not worked since the accident. Michael Richardson, who had commenced work at nine, could recall working three shifts of thirty-six hours duration each, within the space of eleven days. It had not been voluntary overtime. There had been a shortage of boys, and he had been ordered to remain by an overman. He thought all the boys suffered a lot, and he himself suffered from sore arms and legs.

James Richardson, a seventeen year old putter, had been in the mines since he was seven and recalled his first impression of the pit as "a queer place." As a trapper he had been frightened, and was often beaten with a walking stick by the overman. Older boys, engaged in putting and driving, had beaten him with their fists.

Henry Morton, mines agent for the Countess of Durham, said that there was no possible method of working collieries without the employment of very young children. Any law which restricted their employment would immediately produce a scarcity of children and mines would be prevented from "carrying on beneficially." He thought also that "there is a great want of veracity among the boys and all their statements must be received with the greatest caution as they are prone to mischief and insubordination."

He, at least, was honest in his views, unlike another viewer who deprecated the employment of children, while continuing to hire them and disguise their true ages in his returns to the Commissioners.

Commented Leifchild, on the overall situation: "Though very young children are not many in proportion, there are still such a number as is painful to contemplate. The great coalowners will perhaps hear of this for the first time, and I feel a firm belief that they will do so with sorrow and regret. . . ."

He was, of course, being a little over-optimistic.

CHAPTER SEVEN

"*Set Dick the Trapper for some sup,*
We'll suin wesh a' wor faces,
There's neer a lad iv Percy Main,
Be bet this day for five or ten. . . ."
— The Pitman's Luck at the Races.

The job of trapper boy offered embryonic miners their first, harsh taste of life underground. Physical effort was minimal; the harshness of the task lay in the loneliness, the darkness, and the risk of beatings from older boys who were actively involved in the business of coal-getting. It was, however, a job of vital importance to overall mine safety, and a boy who fell asleep could expect no mercy from his superiors.

From the age of around nine the trapper could expect promotion to the position of driver with, probably, responsibility for a pony. This job allowed the boy a certain mobility in the mine, and an increase in wages from 10d. to perhaps 1s. 3d. a day.

The next rung in the ladder of promotion would make him a putter, of which there were several gradations.

The putter was recruited among the older, stronger boys and his task, simply stated, was to heave coal tubs from the coalface to a gathering point, called a landing, or "flats." Putters took pride in their strength, and were inclined to be a race apart in the mine. A "headsman" was a lad not strong enough to "put" alone, who was helped in his task by a younger boy known as a "foal." Where two boys of equal strength were putting, they were known as half-marrows.

Other jobs performed by lads in the pit included that of "craneman" whose task it was to transfer corves and tubs to the outgoing tracks of the mine. There was also a "flatman" who linked the tubs together for the journey to the shaft.

The two hardest tasks in the mine were those of the hewer and the putter. The hewer, by experience, could ease his task by reading the cleat of coal, learning where to place his pick to best effect. He could learn to economise his effort.

Not so the putter. His was a job of sweat-squirting, brutal labour, a task of frenetic tempo and a temper-tearing effort. Their loud-voiced swearing, their inclination to fall to blows at the most minor of pretences, made them a breed to be feared by all. No-one tampered with a putter in the course of his duties. Their ages ranged from the ten or twelve year old "apprentice," or foal, to the hard-muscled eighteen year old, who was on the brink of becoming a fully-fledged hewer.

To the uninitiated visitor, a putter in the full song of labour appeared to be only half-civilised. Tubs rumbled through the underground galleries no more than thirty inches in height. Behind them, black-faced, white-eyed, cursing, was the putter.

Quarrels broke out continuously, especially among the younger boys who, at two to a tub, would dispute the division of effort. The weaker lost the battle – and thereafter would be bullied into performing the major share of the work. It had always been thus. The only technical innovation ever to improve their lot was the invention of iron rail lines which enabled tubs to be put on wheels. It also enabled, said one master, one putter to perform the work previously done by two. The laying-down of underground rail lines earned a rich welcome underground.

Wrote one local poet:

"But heavy puttin's now forgotten,
Sic as we had i' former days,
Ower holly thill and dylls a-splittin',
Trains now a' run on metal ways.
Then bless the man wi' peace and plenty,
That first invented metal plates!
Draw out his years to five times twenty –
And fill his dishes and his plates."

The work of the putters, particularly the young half-marrows and foals, led if anything to even greater abuses than those accorded to trapper boys. Here, indeed, was sweated labour. Just how oppressive it was could not be adequately assessed by the participants because there was no other work for comparison. All carried the mark of their calling in the form of skinned backs, where they had bumped against low roofs. The leather back protectors worn gave cover only to the vital, lower part of the spine (they were hung from the waist) and the shoulders and upper torso took the full brunt of low and uneven roofs. Fully-fledged putters, it was averred, stopped once only when they were down the mine – when they were exhausted. And yet, like troops emerging victorious from battle, they would straighten their backs on the surface and march off home with nonchalant gait. They would arrive on the surface in their working clothes – a vest and short pants – and merely put on their jackets for the walk home. It was an unwritten code among older putters to disdain weather. They never wore greatcoats. At Shilbottle, which supplied the town of Alnwick with its coals, both hewers and putters worked naked because of the heat, and would only put on the working clothes at the end of the shift.

An old hewer, looking back on his career in mining, would chuckle at remembrances of his youthful days as a putter. One such was Thomas Batty who, when interviewed in 1840, was (he thought) ninety-three years of age. This was later verified as correct.

He had been employed at Heaton Colliery until he was eighty-five and said he had enjoyed both good health and fortune largely because of his removal from the coal face to an appointment as an overman when he was forty-five. He recalled starting work in the pits when he was six years old. That was in the Morpeth collieries and his wages were twopence-halfpenny a day. Later, he had become a putter and to earn a wage of ninepence a day had to "put" seventy-two corves. He had transferred to Hartley Colliery where putters were required to work shifts of ten to twelve hours a day. He had then taken up work at Walker where "five groats a day was counted as great wages."

"Putters know nothing now compared to what happened then," he recalled. "There were no safety lamps and no rails . . . just the thill (floor) for the rollies. Boys seldom saw the light in the winter. If a pit was worked on the Friday night until five o'clock, some would go down at eleven o'clock the same night to finish early on Saturday, and then rather than go home they would sleep in the stables."

He had often heard of young drivers working non-stop through three shifts to earn extra money.

He thought that pitwork was healthy providing the mine was kept free of dust and foul air. He himself had suffered from shortness of breath for thirty years and no longer walked ten miles to hear a good sermon as he had done when first married.

The old miner was exercising the prerogative of the aged in emphasising that things were harder in his day for, in truth, little had changed except that the mines had become deeper and – with the opening up of the "gassier" seams – considerably more dangerous.

Certainly, the lot of the putters and drivers seemed just as arduous. At the same colliery where the old man had worked, official returns compiled by colliery officials showed that John Stephenson, "aged about eight," had been employed for two months. It was incorrect. The boy, who was "only three foot eight inches, without shoes, very diminutive, and far from robust or healthy," had been employed below for two years.

He lived in a thatched cottage at Denton, arose each day at 3 a.m. and went down the pit at 4 a.m. He would arrive back home at 5 p.m., have a meal of meat and potatoes, and go directly to bed. He never played. At 3 a.m. the following day he would pick up his "bait" of bread and cheese and set off again to repeat the cycle. He could not read and did not even attend the Sunday School. The putters, he reported, often "brayed" him.

Joseph Stephenson, a ten year old driver, had begun work at nine. For wages of 1s. 3d. per day he would arise at 3 a.m. (except on Saturdays, when he got up at 1 a.m. and finished at 1 p.m.) and start work an hour later. His first task was to get candles to put in his mistress;* he would then go to the stables and yoke his pony. For the ensuing twelve and a half hours he would trundle coal tubs through the mine to the shaft.

"I did not like the pit at first because I was frightened to go in the dark," he said. "I was frightened by a man coming and making fun of me when I was in the dark. He would say 'What are you doing there' and it frightened me and made me cry."

The boy had not yet been injured, although he admitted falling from the rollies at least once daily. On one occasion he escaped injury from runaway rollies by lying flat between the tracks as they passed over him. He could read, but not write, and attended the Methodist Sunday School. His twelve year old brother was a driver in the same mine.

Saunders Blackburn, nine, had been employed at Heaton for a year and was now driving. Already he complained of shortness of breath, for which the colliery doctor had prescribed the treatment of a "blister." The result was six weeks off work. "The pit air does not agree with me," he said.

Another boy, Joseph Mackenzie, often felt sick and suffered from head pains. He regularly regurgitated his food when in the mine. He was also sick each morning at 3 a.m. when he got out of bed to set off for the mine.

Thomas Scattery, a twelve year old, already bore a permanent mark of the mine upon him. He had broken an arm in a rolley accident. It was now permanently bent.

In the low seam collieries it was the putters whose plight gained sympathy. At collieries such as Blaydon Main, where the seam was thirty inches in height, boys of eleven, twelve, and thirteen struggled and strained in conditions described as not much better than sewers. Coal-laden corfs stuck against the roof; young backs were skinned in efforts to free them. Michael Turner, aged fifteen, showed scars on his back as the proof that he was a putter. "My back is strained sore at times," he reported. "The skin is often knocked off. My legs are so tired sometimes I can hardly trail them."

At Shilbottle Colliery the boy putters evolved their own unique method of putting in a thirty inch seam. Twenty boys, aged ten to fourteen, were employed in the work and, to avoid crushing their hands when pushing tubs, they put their heads to the job. An underviewer, Mr. Oliver, told how they took off their hoggers† and stuffed them into their caps to act as cushions.

*Mistress: A tin box containing clay into which a candle was inserted. Used only by drivers.

†A stocking from which the foot was cut off.

Low seams were one hazard for the boy putters; water was another. One fourteen year old half-marrow was obliged to heave tubs through water which reached up to his knees. He passed through the flooded area fifty to sixty times in the course of a shift. Another boy assessed that in the course of a twelve hour shift he would spend upwards of three hours passing through water.

John Dixon, a fourteen year old putter at Percy Main Colliery, about a mile and a halt from North Shields, was another who consistently worked up to the knees in water. It was, he said, a "clean" pit but it had "a vast amount of water in it." He blamed this for giving him head pains. His legs got weak and he often vomited the food he ate below ground, but this had not prevented him staying below ground the previous week for three days and three nights without a stop. "The overman told me to stop down, because John Short and Tom Angus were both ill," he said. Three other putters had gone down the pit on the Thursday at 4 a.m., and had remained until 4 p.m. on the Saturday. The feat had been achieved without benefit of sleep.

Percy Main Colliery – known colloquially as the Howden Pans pit – was to attract a good deal of attention from Government Commissioners. It had been sunk in 1799 when a great deal of trouble had been encountered with quicksands. The bed had been damned back by the use of cast iron tubbing – the first time used in Britain – and the workings were subsequently extended under the bed of the River Tyne. In 1807 it caught fire and the workings were drowned to deal with the outbreak.

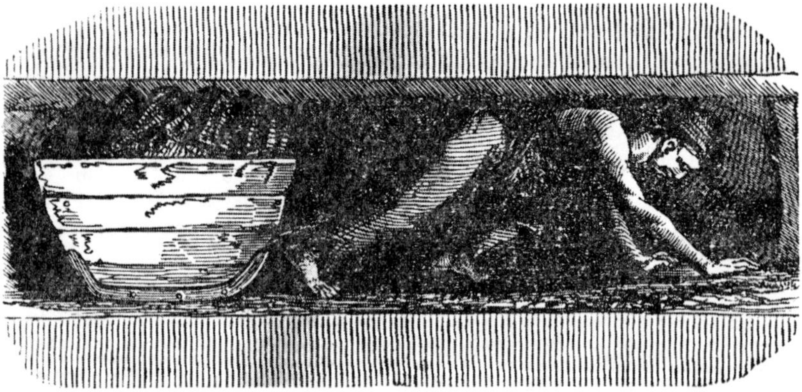

It had then been drained and had since been worked without incident although water continued to be a major problem. It was only kept viable by the use of steam pumping engines totalling 350 horsepower.

What attracted the attention of the Government Commissioners, however, was not so much the problems faced by the colliery owners, Messrs. Humble Lamb and Company, but the conditions faced by their young employees.

Their plight was made known by Luke Gray, a forty-two year old coal hewer. "The boys change places by putting in cavils (a system of drawing lots) once a fortnight. Very little boys are putting as foals and they suffer more than other boys. They are far too young and the work too heavy for them at the age. Men sometimes leave the colliery altogether, because their children are being put too young to this work. These boys are wanted and must be had – they have a great need for boys.

"The extremity of labour is the greatest for foals of any boys. In the low seams, the boys have to travel bent double – and they go about bent double when walking in general. I have seen boys crying in the office because they were bound to be foals."

For its youngest employees the Percy Main Colliery was a place of danger. William Painter, aged twelve, had received treatment from the colliery doctor on five separate occasions for injuries sustained in his job. He had been run over by a train of empty trollies, knocked down by a train of full trollies, fallen and put a nail through his hand, had the spiked handle of his candle lantern through his other hand and, finally, had a rolley roll over his foot. He had, in total, lost over four weeks work.

Taylor Coates, an eleven year old flatman, had twice been injured and missed a total of five weeks work. He was, as a result, permanently lamed and walked with a twisted gait.

Peter Dixon, an eight year old with nine months service at Percy Main had been kicked by a runaway horse and bore permanent facial scars. "The deputies hit me with their sticks sometimes and made me cry," he said.

James Dagliesh, a thirteen year old foal, who had recently worked a twenty-four hour shift non-stop because two other boys had been injured by rollies, reported similar treatment from putters. "They beat me with sticks and make me cry. They hurt me." he said. The boy had been involved in an explosion when he was eleven years old and bore the marks of scorching on his hands and face.

The list of injuries continued, reading like the aftermath of a battle. James Wood, a tiny eight year old, had been taken to the pit by his nine year old brother and had worked below for six weeks. John Fiddle, eight, worked in places where the water was up to the horse's knees; James Carr, a thirteen year old foal, complained that his arms and legs ached because of the work. He said: "I got my hammers (was beaten) twice for being asleep. The putters beat me with sticks and made me cry because I didn't open the door for them. Once I had a bad galloway (horse) and I couldn't make it go, so George Hall, a foal, put some clots or lumps of stuff in my eyes and blinded me for a time. I had to lay idle all that day."

Such was the lot of generations of children in the mines. Was there no way of easing their task?

"Sometimes donkeys are used to assist the putters. At very steep places helpers-up are substituted for these donkeys, and they are much more suitable at slight inclinations. Donkeys are very slow."

That was said by Nicholas Wood, the Killingworth viewer. And he was counted as progressive. . . .

CHAPTER EIGHT

"*We didn't heed much lairnin' then,*
We had ne time for skuel;
Pit laddies wark'd for spendin's syek
And nyen was thowt a fule."

— The Pitman's Happy Times.

The early Victorian mining village in the North East of England was designed, constructed, and operated as an extension of the mine itself. All was subordinate to the pit-head, and welfare of the inhabitants was left to their own provision. Schools were virtually non-existent. The established Church seemed peculiarly unaware of the industrial furore around it, which was odd considering that the Bishop of Durham held and distributed a vast acreage of mining leases, and it was left until the last quarter of the nineteenth century before they embarked on a programme of Church-building. By then, it was almost too late. The civilising influences of Christian teaching came to the mining villages not from the established Church of the land, but from Methodism. It was a do-it-yourself religion which held direct appeal to the untutored miner of the period. From Methodism sprang the Sunday Schools which taught the rudiments of reading and writing, not only to the infant but to whole generations of men who were subsequently to form the resolve that all was not well with the coal mining communities and, if change was needed, it would have to come from the ranks. Thus, it was the "Methodies" who formed the leadership of the union which, in the 1840's, was coming out of embryo in the Northern coalfield.

The miner-Methodists were described as Dissenters, but it was not so. The mining communities, because of Church neglect over the years, had nothing to dissent from. Methodism was to come as the sole religious influence in their lives. Be the chapel ever so humble, it was their own. The chapel might simply consist of a couple of terrace cottages knocked together, and rented to the local Methodist society by an indulgent master, but from it sprang the influences of a better way of life.

The attitude of the masters was one of ambivalence towards the "Methodie" miner. True, in times of dispute, it was the Methodists who led the men. But it was also true that the Methodist miner was a good, conscientious workman, not given to abuse of drink or loss of work.

Generations, however, were to pass before the full impact of the Methodists was to be felt. In the 1840's a pit lad could tell inquiring Commissioners: "In the villages there is a lot of blackguardedness. There are bowling matches, cock fighting, dog fighting and suchlike. Almost all drink sore."

It was assessed that only one in four inhabitants could write their own name, and the annual "bond" or contract, which had to be signed by miners each year, gave proof of it. Such documents were dotted with crosses instead of signatures, with the explanation that "X" was his mark.

A Washington vicar thought that the moral and religious condition of children was particularly low, while the Reverend Hugh Nanny, Vicar of Jarrow, added: "Morals among the children are very bad. Of education there is almost none and their intellect is much debased, both from home habits and the habits of the pit. One great cause is the fact that large families are pent up in small habitations."

The exact figure of illiteracy was difficult to arrive at, short of a complete census, but some indication was drawn from at least one odd source. In the year ending 1840, 141 pitmen were jailed in Newcastle. Of that number, seventy-three were totally illiterate. Sixty-four of them had been imprisoned for breaking the terms of their annual bond to the masters, thirty-two for vagrancy, and four for poaching.

Prior to the Killingworth experiment there was no infants' school, as such, north of the River Wear, and none at all in the south of the coalfield. The only schooling came via the Methodist Sunday Schools, or from ill-attended day schools where the master was perhaps allowed to live in a colliery house on sufferance of the local coalowner. Many were simply disabled workmen; with others, discipline was inclined to be so harsh that children preferred working in the mine. In such schools the whip was held to be the true instrument of education. The "Dame schools" of the era, which supposedly catered for the education of infants, were in fact little more than dumping grounds for children – a place to "get them out of the house for an hour or two." It was not the most laudable of aspirations, but the sheer size of families, and the conditions of living in tiny two room terrace row cottages, made it understandable.

There remained the night schools, which flourished only in the winter months, but attendance was never more than minimal. Few children had the fortitude, or the motivation, to pursue education after working a twelve or fourteen hour shift.

Health, in the villages, was a source of concern to Government investigators. All strangers to the region remarked upon the peculiar physical characteristics of the miner himself. There was the remarkable complexion – likened by one as that of "a pound of wax candles." They were inclined to smallness, with legs bowed and the chest protruding like a pigeon.

It was the physique of degeneration. The Northern miner, with his long tradition of underground work – mining had been carried on long before the industrial revolution – was living proof of the evolutionary theory.

Dr. William Morrison of Chester-le-Street, County Durham, was employed at the collieries owned by the Countess of Durham. The general appearance of the miner was not "healthful," he thought. "His habit is tainted with scrofula. . . . The germ of the physical degeneration may be formed in the youthful days of the pitman. I have never seen a jolly-looking pitman."

Another colliery medic, Dr. T. M. Greenhow, of Walker Colliery, had similar thoughts: "Though the condition of the children, in common with the entire population connected with the collieries, has been considerably improved of late years, many circumstances connected with their mode of life deserve serious consideration and certainly admit of material improvement. The most conspicuous of these is the early age at which they are usally employed in the pit."

Yet a third told how . . . "I have seen children go down here at six years of age and have seen them scarcely able to walk from their tender age and the work."

It was to take the passage of many decades before improved conditions of work and living bettered the physical condition of the Northern miner. The "bantam" battalions of the First World War were well represented in the Durham Light Infantry, and even in the second war military leaders still spoke of the toughness of the "stocky, Northern miner." (In an introduction to a recent history of the Royal Northumberland Fusiliers, Lt. General Sir Brian Horrocks added lustre to the accolade by stating that "Geordies from the northern collieries I have always regarded as making the best infantry in the world.")

CHAPTER NINE

"There's a grand time comin' lads,
A grand time comin'
When collier laddies needn't fret,
For landlords will wor rents forget
I' tha grand time comin'. . . ."

— There's a Grand Time Comin'.

The public outcry which followed the publication of the report from Commissioners inquiring into child employment in the coal mines, brought legislation in 1842 which elevated the age of entry to the mines to ten. It prohibited entirely the employment of women and girls and made provision for the appointment of inspectors to go underground and examine the workings.

The Great Northern coalfield emerged badly from the report, published in two volumes, and was especially condemned for its exploitation of boy labour. It was redeemed only by the discovery that, of all the coalfields, the North was alone in not employing women and girls. The practice had died in the latter half of the eighteenth century, following the death of a woman in a shaft accident at Fatfield, in County Durham.

Legislation was bitterly fought in the House of Commons by the great coalowners of the region, of whom Lord Londonderry was the most loquacious. Government inspectors would, in accordance with the law, enter his pits for inspection purposes but, he reminded Members, there was no law which made it imperative for them to be raised to the surface. Inspectors visiting his collieries could expect to stay down. . . .

The tide of public feeling was against him, however; the Commission's recommendations – mild though they were – became law on 10th August, 1842.

The findings were to form a text book for continental and English reformers.

They told how they were "met at the onset of their labours by the general ignorance which prevailed in reference to the conditions of persons employed in the mining industry into which they were to inquire. It was indeed notorious that in some of these occupations great numbers of children are employed at very early ages. It was equally notorious that some of these occupations are among the most laborious in which it is the lot of human beings to toil.

"But no committee of the House of Commons had been formed to investigate the actual condition of this class, no attempt of any kind had hitherto been made to ascertain the ages at which the children in question begin to work, the number of hours which they labour, the exact nature of such employment on their morals and health."

They found: "That instances occur in which children are taken into these mines to work as early as four years of age, sometimes at five, and between five and six – not infrequently between six and seven; while from eight to nine is the ordinary age at which employment in these mines commences.

"That the great body of the children and young persons employed in these mines are of the families of the adult people engaged in the pits, or belong to the poorest population in the neighbourhood, but in others by the proprietors or contractors. In some there are a small number of parish apprentices who are bound to serve their master until twenty-one years of age, under circumstances of frequent ill-treatment, and under the oppressive condition that they shall receive nothing but food and clothing.

"That the nature of the employment which is assigned to the youngest children, generally that of trapping, requires that they should be in the pit as soon as the work of the day commences; and, according to the present system, that they should not leave the pit before the work of the day is at an end.

"That at different ages, from six years old and upwards, the hard work of pushing and dragging the carriages of coal from the workings to the main ways begins – a labour which all classes of witnesses concur in stating requires the unremitting exertion of all the physical power which the young workers possess."

Turning to the various coalfield areas, the Commission reported that in Northumberland and Durham: "Children begin work at very early ages. One case is recorded in which a child was taken into the pit at four and a half years old."

In the South Durham area "children are sometimes taken down into the pits as early as five years of age, and by no means uncommonly at six." In the same area, "the hours of work in all the well-regulated mines are never allowed to exceed twelve. In many of these collieries formerly, and in some collieries still, when the hewers had cut down what was considered their proper quantity of coal and had departed, the putters, the drivers, and all other persons were expected to remain until the whole of the coals were brought to bank, however long it might be, whether thirteen, fourteen or fifteen or even more hours. If, therefore, any defect occurred in the engine, or any other cause of delay took place, the people were detained below in consequence. Such is not the case now; when the hour comes the sound of "loose" is sent down the shaft, the glad tidings quickly reach the farthest workings, and all come up.

"But this is only true of the well-regulated mines. There are still in the district collieries in which the hours of work are extremely long."

In Northumberland and Durham generally, the regular hours of work were found to be twelve, usually from 4 a.m. to 4 p.m., but "the children and young persons often leave their homes for the pit as early as one or two o'clock in the morning and stay to four o'clock in the afternoon."

In the body of evidence, it was thought undeniable that "the masters might add materially to the physical comforts of the pitman with such facilities as baths and village drainage." It was also reported that the pits had been unmolested by outside influence for so long because the external features of coal country "with rolling smoke, the clatter of strings of coal wagons, is enough to unseat or untenant the seats of the wealthy." The gentry left when pitmen moved in. The absence of resident gentry was to be deplored.

The diminutiveness of the trapper boys was found "remarkable" and this suggested grave doubts of the statements given regarding their ages. That children of five years of age should ever, under any circumstances, be engaged in the pits would have appeared incredible but for the respectability of informants. The trappers themselves had been unable to transmit the bitterness of their thoughts because of their limited vocabulary. It had, for instance, taken Leifchild some time to realise that when a child said it had been "sore tired" it meant that it had actually collapsed with fatigue. A "hurt arm" could mean a fracture.

He himself had been firmly convinced that a limit *had* to be imposed on the ages of children employed in the mines and the employers, in spite of statements to the contrary, could not deem such a measure hostile to the coal trade.

No-one could travel the Northern coalfield and conclude that the labour of children was healthy and beneficial to them (although several viewers thought so in evidence). Children had complained of bad air, vomiting, nausea, boils, sickness, wet working and asthma.

In all, the "radical defects of the district are the early and the prolonged employment of children and young people. These radical defects should be, as speedily as they can be, readily remedied."

The report was presented to the House of Commons by Lord Shaftesbury.

"It is not possible for any man, if he have but a heart in his bosom, to read the details of the awful document without a combined feeling of shame, terror and indignation," he said.

The Act put before the Commons was mild to the point of ineffectuality, but was still considerably modified before it reached its third reading. It was due to coalowner opposition. Its clauses, for instance, were enforceable by fines – but in no case were these to exceed £10.

The provision of inspectors to enforce the legislation proved a particularly bitter pill for coalowners to swallow. In fact, such was their antipathy, that the sole inspector appointed under the Act was unable to fulfil his duty. He never did go down a coal mine in connection with the Act, as he was to confess in his annual report a decade later.

"Everyone would be hostile to such a search. The person attempting it would inevitably be misled and, in two instances where persons attempted it of their own accord, they were maltreated and very nearly lost their lives," he stated in his report for 1854.

It was an incredible comment. The inspector – that "departmental Jack-of-all-trades" H. Seymour Tremenheere – had already confessed in his report for 1851 that the Act was being deliberately flouted. Small boys were, he said, still being taken down the pits in Northumberland and Durham. They were still working shifts of twelve to fourteen hours. The Act had apparently changed little. In 1852 he was reporting the deaths of under-age boys in explosions, and added that the truth about their ages had emerged only because inspectors had attended the inquests.

Public opinion, however, awakened first by the disclosures of the 1842 report, and then by a further series of horrific mine explosions in the following decade, was now firmly against the coalowners and in favour of genuinely active legislation. The result was the passing of a further bill which gained Royal Assent on 14th August, 1850. This enforced the registration of plans, the notification of fatalities, and appointed inspectors with power to enforce the Act by legal proceedings.

By February, 1860, the age limit for the employment of boys had been raised to twelve (with the exception of those who could read and write).

The appointment of the inspector under the 1842 Act was finally lapsed and his duties handed over to safety inspectors. By 1872 the hours worked by boys was restricted to ten hours with the provision that they were to attend school not less than twenty hours each fortnight.

The "good times" had finally arrived. . . .

CHAPTER TEN

"Huz colliers, for a' they can say,
Hez byeth heeds and hearts that are sound –
And if we're but tyun in wor way,
There's few better cheps above ground."

— The Collier's Keek at the Nation.

Improved education and working conditions, the agitation for an extension of the franchise, plus the growing strength of trades unionism were to bring to the fore in Victorian years a number of remarkable men who all began their working lives as child miners. Men such as Thomas Burt, the "first working-man Member of Parliament," and John Wilson, M.P., leader of the Durham Miners' Union, both wrote autobiographies and were to recall with bitterness the conditions which prevailed in their youth. Another was George Parkinson, a patriarch of Durham Methodism, who was born in 1828 and began work as a trapper when he was nine. Sundays, he recalled in later life, were regarded in the pit villages as the jewels of the week. Sixty years later he was still able to recollect with absolute clarity an incident which illustrated this. After six successive days of arising at 3 a.m. for his fifteen hour shifts, he had gone to bed on the Saturday night in a state of exhaustion (he was just over nine years of age). At 5 a.m. he awoke, but could not find his pit clothes by the fire. "I was just getting peevish, when I saw my Sunday clothes laid on the chair beside the bed. It dawned on me that the day was Sunday, and with feelings of joy and satisfaction such as no words could express, I sprang back into bed and drifted off to sweet sleep. . . ."

In his first day in the mine he had been taken below by his father. At the shaft bottom, a candle set in a "sticking" of clay was put in his hands and he was led to a trap door. His father had then hewn out a recess in the gallery wall where the new trapper boy could sit comfortably. With the admonition to "keep thi' heart up, hinney; thoo'll mak' a good pitman yet," he had left the young Parkinson to his job.

Later, Parkinson was involved in an explosion and narrowly escaped with his life. It coincided with an opportunity to leave the pits, and he accepted with alacrity.

John Wilson, M.P., was a child when his mother died, and he had subsequently travelled the length of Britain on his father's shoulders, in a quest for work on the railways then thrusting through the country-side.

In 1849 both had obtained work in the quarries at Stanhope. There, his father died suddenly of cholera, nursed only by his son because local people were afraid of infection. By various routes, Wilson found his way to the coastal collieries.

"It was the unquestioned prerogative of the official to punish when he thought fit," wrote Wilson. He recalled an incident when putters at Ludworth Colliery, near Durham, had a dispute over money and decided to leave the pit. En route, they were met by an overman. The sight was enough to break the resolve of the "striking putters" who fled back to the workings, but Wilson, the youngest, stood his ground.

"He knocked me down into the middle of the way, and kicked me for some time. Then he asked again if I would go back. I answered 'You may kill me if you want, but I am not going in.' What would have been the consequence I cannot say, for just then a hewer named Quinn came out, and he stopped him and said, 'Let the lad alone, you brute, or it will be you and me for it.' I was in bed for three days."

Tommy Burt began work at Haswell Colliery, aged ten, just a year or two after an explosion there which had claimed ninety-five lives.

His first task was that of trapper. "The dullness and loneliness were oppressive," he said. "A shy, sensitive boy, I was thrown – or, rather, had voluntarily rushed – into fortune's way to take my part as best I could, in the rough conflicts of daily pit life.

"The clatter, the bustle and confusion, the darkness relieved only by the glimmering, flickering lamps, the men and boys flitting hither and thither, resembling grim shows – all this seemed a fantastic dream. . . ."

It was pony putting days which left their mark on Burt's memory and he was to recall them as "the most miserable days of my life." He worked shifts of not less than fourteen or fifteen hours and the strain imposed upon him was beyond his strength. His father had argued with the colliery manager that young Burt was too young for the task. The reply was: "He's big enough and to that work he must go."

Wrote Burt: "I suffered severely at the time, and, possibly, consequences followed which affected me in later years. Those weary loads under which I was called upon to sweat and groan in my boyhood, not improbably tended to arrest my growth, and to render me for the rest of my life less healthy and vigorous than I should otherwise have been."

It was as a putter that Burt learned to swear. The work left him so utterly used-up that often he would return home and throw himself unwashed on the floor and fall asleep. Only with difficulty could he be pursuaded to remove his pit clothes for bed. He would plead in vain to be allowed to lie in his clothes so that he would be ready for work the following day. This was at Murton Colliery in County Durham, a place where – said Burt – he had never before seen "so many crutches, so many empty jacket sleeves, so many wooden legs."

Burt, Wilson and Parkinson were not alone in overcoming the disadvantages of ill-used youth in the mines. Others surmounted all difficulty to win recognition in various fields of endeavour, but all were to carry memories of their early pit days to the grave. In life, they could only look back at their own childhood experiences with a sense of disbelief.

THE ROMAN WALL

All these books are lavishly illustrated

HADRIAN'S WALL. Guide to the Central Sector by *Robin Birley*. Sixty four pages. **60p.**

THE ROMANS IN NORTH BRITAIN by *T. H. Rowland*. Sixty-four pages. **80p.**

THE ARMY OF HADRIAN'S WALL by *Brian Dobson* and *David Breeze*. Forty-eight pages. **60p.**

BUILDING THE ROMAN WALL by *Brian Dobson* and *David Breeze*. Revised and with new illustrations. **40p.**

SHORT GUIDE TO THE ROMAN WALL. Fifty-six pages. A detailed and up to date guide with numerous maps. Remarkable value. By *T. H. Rowland*. Illustrated in colour. **60p.**

DISCOVERIES AT VINDOLANDA. 1973. This illustrated booklet describes many of the remarkable finds made at Vindolanda during recent months. **50p.**

HOUSESTEADS ROMAN FORT by *R. Birley*. Four pages in colour. **35p.**

CHARTS OF THE ROMAN WALL. A series of magnificent coloured reconstructions of the Roman Wall. Size seventeen inches by twenty-five inches. Price **60p** each.

1. **South Gate, Housesteads.**
2. **Mansio or Inn at Vindolanda.**
3. **Latrines at Housesteads.**

VINDOLANDA IN COLOUR. A magnificent souvenir of the Roman fort and settlement which has attracted so much attention recently. **80p**

THE ROMAN WALL RECONSTRUCTED. Twelve magnificent paintings in full colour showing the Roman Wall as it was when built. Paintings by *R. Embleton*. Text by *C. M. Daniels*. **80p.**

DERE STREET. Roman Road North by *T. H. Rowland*. From York to Scotland. **60p.**

From Peter Anthony's column in "The Journal" 17th September, 1976

It must have been a unique occasion. A civic reception in Newcastle's stately Mansion House – for a book.

"It's certainly the first time a Lord Mayor of this City has got himself involved in the bookselling business", said the Lord Mayor, Councillor Hugh White.

It was a unique book, of course: *Historic Newcastle*, five poundsworth of sheer delight for anyone who likes to escape into the past. And for its author, Frank Graham, a very special milestone in his 10 years in the publishing business. Those 10 years have been largely devoted to promoting the North-East and one of the results has been to provide Newcastle with more guide books and volumes of historic interest than probably any place outside London.

Frank has taken a special interest in his latest publication. For years he has been aggravated to see contracts for quality colour printing going outside this county. His claim that "we can do it better here" seems to be borne out by *Historic Newcastle*, an absorbing collection of illustrations accompanied by a well-researched text. Thirty-three of the plates are in full colour and the whole production process – writing, printing, book-binding – was done within a mile or so of the Mansion House.

Arousing special interest at the civic launching was a painting by Ralph Hedley, an unsung local artist working at the turn of the century. It is of a Newcastle newsboy, reproduced in colour for the first time. A companion picture by the same artist now holds a place of honour among the Mansion House's treasures.

"All the illustrations were chosen specially to draw attention to the city", Frank told me as he autographed a copy of his book.

HISTORIC NEWCASTLE

PRICE £5

Historic Newcastle is one of the finest of our publications. We have now issued 300 local books in two million copies and our present venture is intended as a fitting climax to such a varied effort to promote the history and culture of Northumbria. We hope the £12,000 spent on this book will be appreciated by those who study its contents. *Historic Newcastle* tries to present the history of our own city through the medium of illustrations, 33 of which are in colour. It is not fully comprehensive, since the history of Newcastle is so varied and important, but few of the memorials of our past greatness have been omitted from the 160 paintings and engravings here illustrated.

THE OLD HALLS, HOUSES AND INNS OF NORTHUMBERLAND

This is the first comprehensive book on the Halls and Houses of Northumberland to be published. It is a companion volume to our *Castles of Northumberland* published last year. 312 pages. 281 Houses. 301 Illustrations.

PRICE £6

Brinkheugh

"It yields 301 buikdings for Frank Graham's alphabetical description, all rooted in rich historical soil: is that still possible anywhere else in England. What in any case would be unlikely elsewhere is a record so visual with 281 illustrations from prints or photographs. Frank Graham as a publisher, as well as author, is passionately concerned about illustrating his books and this companion to his 'Castles of Northumberland' completes a splendid pair".

Northern Echo